Unlocking the Secrets of Black Holes A Journey into Astrophysics

Hansen

Copyright © [2023]

Title: Unlocking the Secrets of Black Holes A Journey into Astrophysics

Author's: Hansen.

This book was printed and published by [Publisher's: Hansen]   in [2023]

ISBN:

TABLE OF CONTENTS

Chapter 1: Introduction to Black Hole Astrophysics

The Fascination with Black Holes

Black holes have been captivating the human imagination for centuries. These mysterious cosmic entities have served as a source of inspiration for scientists, philosophers, and science fiction writers alike. In this subchapter, we will delve into the fascination that black holes hold over our collective consciousness and explore the reasons behind their allure.

One of the primary reasons for the fascination with black holes is their incredible and mind-boggling properties. These celestial objects are formed from the remnants of massive stars that have collapsed under their own gravity. The sheer density of a black hole is unfathomable, with all its mass concentrated into an infinitesimally small point called a singularity. The gravitational pull exerted by a black hole is so immense that nothing, not even light, can escape its clutches once it crosses the event horizon.

This concept alone challenges our understanding of the laws of physics and prompts us to question the very fabric of the universe. The enigmatic nature of black holes, their ability to distort space and time, and the possibility of wormholes and parallel universes within them, all contribute to their allure.

The implications of black holes extend beyond astrophysics and into the realms of philosophy and existentialism. They represent the ultimate unknown, the great void in our knowledge. They force us to ponder the nature of existence, the fate of matter, and the limits of our understanding. The mystery of what happens beyond the event horizon tantalizes our curiosity and pushes us to explore the boundaries of human knowledge and imagination.

Furthermore, black holes offer a unique laboratory for studying the fundamental laws of physics. They provide opportunities to test Einstein's theory of general relativity and investigate the nature of spacetime. By analyzing the effects of black holes on surrounding matter and their influence on the cosmic landscape, scientists can gain insights into the fundamental nature of the universe.

In conclusion, the fascination with black holes is rooted in the awe-inspiring properties they possess, the existential questions they raise, and the scientific opportunities they present. By unlocking the secrets of black holes, we push the boundaries of our understanding and embark on a journey of discovery that captivates not only astrophysicists but also every curious individual seeking to comprehend the wonders of the cosmos.

Brief History of Black Hole Study

Since their theoretical prediction in the early 20th century, black holes have captivated the minds of scientists and the general public alike. This subchapter will take you on a journey through the fascinating history of black hole study, showcasing the key milestones and breakthroughs that have shaped our understanding of these enigmatic cosmic entities.

The concept of black holes was first proposed by Albert Einstein's theory of general relativity in 1915. However, it wasn't until the 1930s that Indian astrophysicist Subrahmanyan Chandrasekhar introduced the idea of a maximum mass limit for white dwarf stars, leading to the concept of black holes as the endpoint of stellar evolution. This pivotal breakthrough laid the foundation for further exploration and research.

In the 1960s, the renowned physicist Stephen Hawking made groundbreaking contributions to our understanding of black holes. Through his work on Hawking radiation, he theorized that black holes are not entirely black but emit particles, eventually evaporating over time. This discovery challenged established notions and opened up new avenues of research, placing black holes at the forefront of astrophysical investigations.

The study of black holes gained further momentum with the advent of space-based observations. In 1971, the first X-ray observatory, known as the Uhuru satellite, was launched, allowing scientists to detect and analyze black hole candidates through the X-ray emissions produced by their accretion disks. This marked a significant step forward in the identification and characterization of black holes.

The 1990s witnessed a revolution in black hole research with the launch of the Hubble Space Telescope. This pioneering instrument provided unprecedented high-resolution images, enabling scientists to observe the dynamic interplay between black holes and their surrounding galaxies. Additionally, advancements in computer simulations and numerical modeling allowed researchers to simulate the behavior of matter falling into black holes, shedding light on the intricate processes at work.

In recent years, the field of black hole study has experienced remarkable progress. The first-ever direct image of a black hole's event horizon, captured by the Event Horizon Telescope in 2019, was a monumental achievement. This image provided visual confirmation of the existence of black holes and offered valuable insights into their nature and behavior.

Today, black hole research continues to push the boundaries of astrophysics. Scientists are exploring the connection between black holes and other cosmic phenomena, such as gravitational waves and dark matter. With each new discovery, our understanding of these mysterious entities deepens, unraveling the secrets of the universe on a grand scale.

In conclusion, the study of black holes has evolved significantly over the years, thanks to the contributions of brilliant scientists and technological advancements. As we delve deeper into the mysteries of these cosmic giants, we are unlocking the secrets of the universe and expanding our knowledge of the fundamental forces that shape the cosmos.

Theories and Concepts in Black Hole Astrophysics

Black holes have captivated the imaginations of scientists and the general public alike for decades. These enigmatic cosmic phenomena, characterized by their incredibly strong gravitational pull, continue to intrigue and mystify us. In this subchapter, we explore the various theories and concepts in black hole astrophysics that have emerged through extensive research and observations.

One of the fundamental theories in black hole astrophysics is General Relativity, proposed by Albert Einstein. According to this theory, the presence of a massive object, such as a black hole, warps the fabric of spacetime, creating a gravitational field from which nothing, not even light, can escape. This concept forms the basis for our understanding of black holes and their behavior.

Another important concept in black hole astrophysics is the event horizon, the boundary beyond which nothing can escape the gravitational pull of a black hole. Once an object crosses this boundary, known as the point of no return, it is irretrievably drawn into the black hole's singularity, a region of infinite density. The event horizon plays a crucial role in defining the observable properties of black holes.

Black hole accretion is another fascinating concept in astrophysics. When matter, such as gas or dust, falls towards a black hole, it forms an accretion disk around it. This disk heats up due to friction and emits powerful radiation, making it observable through various telescopes and instruments. The study of accretion disks has provided valuable insights into the behavior and properties of black holes.

Scientists have also proposed the existence of supermassive black holes at the centers of galaxies. These behemoths, with masses millions or even billions of times that of our Sun, are thought to play a crucial role in the formation and evolution of galaxies. The study of these supermassive black holes has opened up new avenues of research and deepened our understanding of the universe.

In recent years, the field of black hole astrophysics has witnessed significant advancements, thanks to groundbreaking observations and theoretical breakthroughs. These include the direct imaging of a black hole's shadow, the detection of gravitational waves from black hole mergers, and the exploration of black hole thermodynamics. These developments have not only confirmed existing theories but have also posed new questions, driving further research and exploration.

In conclusion, the study of black hole astrophysics encompasses various theories and concepts that continue to push the boundaries of our knowledge. From General Relativity to accretion disks, event horizons to supermassive black holes, each discovery brings us closer to unlocking the secrets of these cosmic marvels. As we delve deeper into the mysteries of black holes, we gain a better understanding of the universe and our place within it.

Chapter 2: What is a Black Hole?

Understanding Gravity and General Relativity

Gravity is one of the fundamental forces in the universe. It not only keeps us grounded on Earth but also governs the motion of celestial bodies and shapes the very fabric of space and time. In the subchapter "Understanding Gravity and General Relativity," we will delve into the intricacies of this force and explore the revolutionary theory of General Relativity proposed by Albert Einstein.

Gravity, as we commonly perceive it, is the force that attracts objects towards each other. However, the true nature of gravity goes beyond this simple concept. According to General Relativity, gravity is the curvature of spacetime caused by the presence of mass and energy. It suggests that massive objects, such as stars and planets, create a dent in the fabric of spacetime, and other objects are compelled to follow the curved path around them.

To comprehend this mind-bending theory, let's consider the famous analogy of a rubber sheet. Imagine a flat rubber sheet representing the fabric of spacetime. When a massive object, like a bowling ball, is placed on the sheet, it creates a dip or curvature. Now, if you roll a smaller object, like a marble, nearby, it will follow a curved path around the bowling ball. This is analogous to how gravity works in the universe, where the massive objects create curves in spacetime, and other objects are influenced by this curvature to move along specific paths.

General Relativity also predicts the existence of black holes, which are the most extreme manifestation of gravity. These mysterious entities are formed when massive stars collapse under their own gravity, creating an incredibly dense region where gravity becomes infinitely strong. The gravitational pull of a black hole is so intense that nothing, not even light, can escape its clutches. They are like cosmic vacuum cleaners, devouring anything that comes too close.

Understanding gravity and General Relativity is crucial for astrophysics as it forms the foundation for studying black holes, the evolution of the universe, and the behavior of galaxies. It allows us to unravel the secrets of these cosmic phenomena and gain insights into the nature of space, time, and the universe itself.

In the next sections, we will explore the experimental evidence supporting General Relativity, such as the bending of starlight around massive objects and the detection of gravitational waves. We will also discuss the implications of this theory on our understanding of the universe and the ongoing research to unveil the mysteries of black holes.

Whether you are a curious reader with a fascination for astrophysics or a student aiming to delve deeper into this captivating field, "Understanding Gravity and General Relativity" will open the door to a realm of knowledge that will transform your perception of the universe.

Singularity, Event Horizon, and Schwarzschild Radius

In the vast expanse of the cosmos, there exist celestial objects that challenge our understanding of the universe - black holes. These enigmatic entities have captivated the minds of scientists and the public alike, sparking curiosity and awe. In this subchapter, we delve into three fundamental concepts related to black holes - Singularity, Event Horizon, and Schwarzschild Radius - shedding light on their mysteries and unveiling the secrets they hold.

At the heart of a black hole lies the Singularity, a point of infinite density and gravitational pull. It is a region where the laws of physics as we know them break down, and our current understanding fails to provide a complete description. Within the Singularity, matter is crushed to an unimaginable density, and space-time itself becomes warped beyond comprehension. It is a true cosmic enigma that challenges our knowledge of the universe.

Surrounding the Singularity is the Event Horizon, a boundary beyond which nothing, not even light, can escape the gravitational grip of the black hole. This invisible threshold marks the point of no return, where the gravitational pull becomes so intense that even the fastest objects in the universe are unable to break free. Once an object crosses the Event Horizon, it is forever trapped within the clutches of the black hole, disappearing from our observable universe.

The Schwarzschild Radius, named after the German physicist Karl Schwarzschild, is a mathematical concept that defines the size of the Event Horizon. It represents the distance from the center of the black hole to the point where the gravitational pull becomes overwhelming. Any object within this radius is inevitably bound for the Singularity. The Schwarzschild Radius is directly proportional to the mass of the black hole, meaning that more massive black holes have larger Event Horizons.

Understanding these concepts is crucial in unraveling the mysteries of black holes and their role in the cosmos. They challenge our understanding of physics, bending the fabric of space and time, and providing a glimpse into the extreme environments that exist in the universe. By studying black holes, astrophysicists strive to unlock the secrets of the universe, exploring the nature of gravity, the birth and death of stars, and the ultimate fate of our universe itself.

In conclusion, the concepts of Singularity, Event Horizon, and Schwarzschild Radius are fundamental to comprehending the nature of black holes. They represent the extreme limits of our understanding, pushing the boundaries of astrophysics and challenging the very fabric of reality. By unlocking the secrets of black holes, we embark on a journey into the unknown, expanding our knowledge of the cosmos and our place within it.

Types of Black Holes: Stellar, Supermassive, and Intermediate

One of the most intriguing aspects of black holes is the diversity they exhibit in terms of their size and mass. In the vast expanse of the universe, black holes come in three main types: stellar, supermassive, and intermediate. Each type has its own unique characteristics and plays a significant role in the study of astrophysics.

Stellar black holes are the smallest of the three types, yet they possess an immense gravitational pull. They are formed from the remnants of massive stars that have undergone a supernova explosion. When a star exhausts its nuclear fuel, it collapses under its own gravitational force, compressing its mass into an incredibly small space. This compression creates a singularity, a point of infinite density, and forms a stellar black hole. These black holes typically have a mass of a few times that of our sun, but their size is incredibly small, with a diameter of only a few kilometers. Stellar black holes are fascinating to astrophysicists as they serve as a window into the life cycle of massive stars.

On the other end of the spectrum, we have supermassive black holes, which are millions or even billions of times more massive than our sun. These colossal entities reside at the centers of galaxies, including our very own Milky Way. The origins of supermassive black holes remain a mystery, but it is believed that they grow over time by devouring surrounding matter and merging with other black holes. Their immense gravitational pull affects the dynamics of entire galaxies, influencing the formation of stars and the distribution of matter. Understanding supermassive black holes is crucial in unraveling the mysteries of galaxy formation and evolution.

Finally, we have intermediate black holes, which fall between the size range of stellar and supermassive black holes. These black holes have a mass ranging from hundreds to thousands of times that of our sun. Intermediate black holes are relatively rare, and their formation mechanisms are still not entirely understood. However, they provide valuable insights into the transition between stellar and supermassive black holes, serving as a missing link in our understanding of black hole evolution.

In conclusion, the study of black holes encompasses a wide range of sizes and masses. Stellar, supermassive, and intermediate black holes each have their own distinct characteristics and significance in the field of astrophysics. By exploring these different types, scientists can unlock the secrets of the universe and gain a deeper understanding of the fundamental forces that shape our cosmic environment.

Chapter 3: Formation and Evolution of Black Holes

Stellar Evolution and Supernovas

In the vast expanse of the universe, stars are born, live out their lives, and eventually meet their spectacular end. This fascinating process, known as stellar evolution, holds the key to understanding the birth of black holes, those enigmatic cosmic entities that captivate our imagination. In this subchapter, we will embark on a journey into the realm of astrophysics to unlock the secrets of stellar evolution and supernovas.

Stars are born from vast clouds of gas and dust called nebulae. These nebulae collapse under their own gravity, triggering the birth of new stars. As the core of a newborn star ignites, nuclear fusion begins, converting hydrogen into helium and releasing a tremendous amount of energy. This energy counteracts the gravitational force, creating a delicate balance that sustains the star's life.

As a star ages, it undergoes various stages, transforming gradually. Small- to medium-sized stars, like our Sun, become red giants as they exhaust their hydrogen fuel. During this phase, the star swells up and engulfs any nearby planets, leaving behind a white dwarf, a dense remnant of the star's core.

However, more massive stars follow a different path. When they run out of hydrogen, their core contracts, causing a rapid fusion of helium into heavier elements like carbon, oxygen, and eventually iron. This process releases an immense amount of energy, leading to a cataclysmic explosion known as a supernova.

Supernovas are some of the most powerful events in the universe. They can outshine entire galaxies and release more energy in a few weeks than our Sun will emit over its entire lifetime. During a supernova, the outer layers of the star are blasted into space, scattering heavy elements across the cosmos. These elements become the building blocks for future generations of stars and planets.

But what remains after a supernova? For stars with masses at least three times that of the Sun, the core collapses to form a black hole. These gravitational behemoths have an incredibly strong gravitational pull that not even light can escape from, earning them their name.

Understanding the intricacies of stellar evolution and supernovas allows us to comprehend the origins of black holes, those enigmatic and powerful entities. By studying the life cycles of stars, astrophysicists can delve deeper into the mysteries of the universe, uncovering the secrets of black holes and the fundamental laws that govern our cosmos.

So, join us on this enthralling journey into astrophysics as we explore the fascinating processes of stellar evolution and supernovas, unraveling the secrets of black holes and gaining a deeper understanding of our place in the vastness of the universe.

The Birth of Black Holes

In the vast expanse of our universe, where stars twinkle and galaxies dance, lies one of the most intriguing and enigmatic celestial phenomena - the black hole. These cosmic monsters have fascinated scientists and astronomers for centuries, captivating our imagination and challenging our understanding of the universe. In this subchapter, we will embark on a journey into the birth of black holes, unraveling the secrets that shroud these mysterious entities.

Black holes are born from the remnants of massive stars, whose core collapses under its own gravity after exhausting its nuclear fuel. This cataclysmic event, known as a supernova, marks the beginning of the black hole's existence. As the star collapses, it forms an incredibly dense region in space, known as a singularity, where gravity becomes infinitely strong. This singularity is surrounded by a boundary called the event horizon, beyond which nothing, not even light, can escape its gravitational pull.

Understanding the birth of black holes requires delving into the fundamental principles of astrophysics. When a star several times more massive than our Sun exhausts its nuclear fuel, it can no longer support its own weight, leading to a rapid collapse. The immense gravitational forces cause the star to shrink into a fraction of its original size, while its mass remains intact, resulting in a highly dense object.

During this collapse, a tremendous burst of energy is released, creating a supernova explosion that can outshine an entire galaxy. The outer layers of the star are expelled into space, enriching the surrounding interstellar medium with heavy elements essential for the formation of new stars and planetary systems.

Meanwhile, the core of the collapsing star continues to compress, becoming denser and denser. At a certain critical point, the gravitational pull becomes so intense that not even light can escape its clutches. This point marks the birth of a black hole - a region of space where gravity has triumphed over all other forces.

The birth of black holes is a mesmerizing yet violent process that shapes the cosmos. These celestial powerhouses, with their intense gravitational pull, have a profound impact on their surroundings, shaping the evolution of galaxies and influencing the formation of new stars.

As we delve deeper into the secrets of black holes, we will explore their fascinating properties, the mind-bending physics behind them, and the role they play in shaping the universe. Join us on this journey of discovery, where we unlock the secrets of black holes and venture into the unknown reaches of astrophysics. Whether you're an avid astrophysics enthusiast or simply curious about the wonders of the universe, this subchapter will take you on an awe-inspiring exploration of the birth of black holes.

Growth and Mergers of Black Holes

One of the most intriguing and enigmatic phenomena in the universe is the growth and mergers of black holes. These cosmic entities, which were once considered mere theoretical constructs, have now become the focus of intense study and fascination within the field of astrophysics. In this subchapter, we will delve into the captivating world of black hole growth and mergers, unlocking the secrets that lie within.

Black holes are formed from the remnants of massive stars that have exhausted their nuclear fuel and undergone a catastrophic collapse. As matter collapses under its own gravitational pull, a singularity is formed, surrounded by an event horizon beyond which nothing, not even light, can escape. While this process of black hole formation is captivating in itself, what truly astounds scientists is the subsequent growth and mergers of these celestial entities.

Black holes can grow in two primary ways: through the accretion of matter and through mergers with other black holes. Accretion occurs when a black hole attracts and engulfs nearby matter, such as gas, dust, or even other stars. As this matter spirals into the black hole's event horizon, it releases an enormous amount of energy in the form of intense radiation and jets of plasma, creating some of the most luminous objects in the universe, known as active galactic nuclei.

In addition to accretion, black holes can also merge with one another. When two black holes are in close proximity, their immense gravitational forces can cause them to spiral towards each other. As they approach, the black holes release gravitational waves – ripples in the fabric of spacetime – which carry away energy and angular momentum. Eventually, the two black holes merge into a single, more massive black hole, releasing an immense amount of gravitational wave energy in the process.

The study of black hole growth and mergers provides invaluable insights into fundamental astrophysical processes and the nature of gravity itself. By observing and analyzing the electromagnetic and gravitational wave signatures associated with these events, scientists can better understand the dynamics of black hole accretion disks, the formation and evolution of galaxies, and the properties of spacetime itself.

In conclusion, the growth and mergers of black holes are awe-inspiring phenomena that continue to captivate scientists and astrophysics enthusiasts alike. Through the process of accretion and mergers, black holes shape the very fabric of our universe, leaving behind a trail of gravitational waves and radiation that carry the secrets of our cosmos. By unlocking these secrets, we inch closer to understanding the mysteries of black holes and the universe at large.

Chapter 4: Observing Black Holes from Earth

Telescopes and Observatories

In the vast expanse of the cosmos, telescopes act as our windows into the mysteries of the universe. These remarkable instruments offer a glimpse into the depths of space, allowing us to explore celestial bodies, study distant galaxies, and unlock the secrets of black holes. In this subchapter, we will delve into the fascinating world of telescopes and observatories, and how they have revolutionized the field of astrophysics.

Telescopes come in various shapes and sizes, each designed to capture different wavelengths of light. From the visible spectrum to X-rays and radio waves, these instruments allow us to observe the universe in ways that our eyes alone cannot comprehend. Whether it's the Hubble Space Telescope orbiting high above the Earth, or ground-based observatories like the Keck Observatory in Hawaii, each telescope plays a crucial role in expanding our knowledge of astrophysics.

Observatories, on the other hand, are the homes of telescopes. These scientific facilities are strategically located in remote areas, away from light pollution, to ensure the clearest possible observations. Observatories can be found in deserts, mountaintops, and even deep underground. They provide scientists with the ideal conditions to peer into the depths of space and unravel the mysteries of black holes.

The development of telescopes and observatories has revolutionized astrophysics, enabling us to make groundbreaking discoveries. Through these powerful instruments, we have observed the birth and death of stars, detected distant exoplanets, and even witnessed the collision of black holes. With each new observation, we gain a deeper understanding of the universe and our place within it.

Moreover, telescopes and observatories have paved the way for technological advancements. From the invention of the first refracting telescope by Galileo Galilei to the cutting-edge adaptive optics used today, these instruments have constantly evolved and improved. They have pushed the boundaries of what we thought was possible and continue to do so, opening up new avenues for scientific exploration.

In conclusion, telescopes and observatories are the backbone of astrophysics. They allow us to peer into the depths of space, uncovering the secrets of black holes and unraveling the mysteries of the universe. These incredible instruments have not only expanded our understanding of astrophysics but have also driven technological advancements. As we continue to invest in these scientific marvels, we can look forward to even more astonishing discoveries and a deeper appreciation of the vastness and complexity of our universe.

Studying Black Holes through Electromagnetic Radiation

One of the most intriguing and mysterious phenomena in the universe is the black hole. These enigmatic cosmic entities have fascinated scientists and astrophysicists for decades, and the quest to understand them continues to captivate our imagination. One of the ways we unlock the secrets of black holes is through the analysis of electromagnetic radiation.

Electromagnetic radiation is a form of energy that encompasses a wide range of wavelengths, including visible light, radio waves, X-rays, and gamma rays. By studying the electromagnetic radiation emitted by black holes, scientists can gather valuable information about their properties, behavior, and effects on their surroundings.

One of the key aspects of studying black holes through electromagnetic radiation is observing the accretion disks that form around them. When matter falls into a black hole, it spirals inward, forming a disk of superheated gas and dust. This disk emits intense amounts of electromagnetic radiation across the entire spectrum. By analyzing this radiation, astrophysicists can determine the temperature, density, and composition of the accretion disk, providing insights into the feeding habits and growth of black holes.

Furthermore, black holes often exhibit powerful jets of high-energy particles that are accelerated to nearly the speed of light. These jets emit intense radio waves and X-rays, which can be detected and analyzed by advanced observatories such as radio telescopes and X-ray satellites. By studying the properties of these jets, scientists can gain a better understanding of the mechanisms behind their formation and the immense energies involved.

Additionally, electromagnetic radiation allows us to investigate the gravitational effects of black holes on their surroundings. When a black hole interacts with a nearby star, it can cause the star to emit intense bursts of X-rays. By studying these X-ray emissions, scientists can determine the mass and spin of the black hole, as well as explore the dynamics of the binary system it forms with the star.

In conclusion, electromagnetic radiation plays a crucial role in unraveling the mysteries of black holes. Through the analysis of various wavelengths, astrophysicists can investigate the accretion disks, jets, and gravitational effects associated with these cosmic enigmas. The insights gained from studying black holes through electromagnetic radiation not only deepen our understanding of these extraordinary objects but also shed light on the fundamental laws of physics that govern the universe.

Detecting Black Holes through Gravitational Waves

In the fascinating realm of astrophysics, black holes have long captured the imagination of scientists and the general public alike. These enigmatic cosmic entities, with their immense gravitational pull, have the power to bend the fabric of space and time, creating a gravitational force from which not even light can escape. But how do we detect something that emits no light?

Enter gravitational waves, the ripples in spacetime that propagate outward from massive objects, including black holes. These waves, first predicted by Albert Einstein over a century ago in his theory of general relativity, provide a unique tool for studying the elusive black holes.

Gravitational wave detectors, such as the Laser Interferometer Gravitational-Wave Observatory (LIGO), have revolutionized our ability to observe the universe. By observing the tiny disturbances caused by gravitational waves passing through Earth, scientists can now detect the mergers of black holes millions or even billions of light-years away.

When black holes merge, they create a cataclysmic event that releases an enormous amount of energy in the form of gravitational waves. These waves carry vital information about the black holes themselves, including their masses, spins, and distances from Earth. By carefully analyzing the signal received by gravitational wave detectors, astrophysicists can decipher the intricate dance of black holes as they spiral towards each other and eventually merge into a single entity.

The detection of gravitational waves from black hole mergers has opened up a new window into the study of these mysterious cosmic objects. Not only does it confirm the existence of black holes, but it also provides crucial insights into their formation and evolution. By studying the properties of black hole mergers, scientists can gain a deeper understanding of the mechanisms behind their formation, the environments in which they reside, and the role they play in shaping the cosmos.

The detection of gravitational waves has also allowed scientists to explore the boundaries of general relativity, Einstein's theory of gravity. By comparing the observed gravitational wave signals with the predictions of general relativity, researchers can test the limits of our current understanding of the universe and potentially uncover new physics.

In conclusion, the ability to detect black holes through gravitational waves has revolutionized the field of astrophysics. It has provided us with a new way to observe and study these enigmatic cosmic entities, shedding light on their properties and expanding our knowledge of the universe. Gravitational wave detectors continue to unlock the secrets of black holes, captivating both scientists and enthusiasts on a journey into the depths of astrophysics.

Chapter 5: Black Hole Dynamics and Mechanics

Accretion Discs and the Innermost Stable Circular Orbit

In the vast expanse of space, where gravity reigns supreme, lies one of the most intriguing and enigmatic phenomena in the universe – black holes. These celestial objects possess a gravitational pull so strong that nothing, not even light, can escape their grasp. But how do black holes form, and what happens to matter that falls into them? The answer lies in the concept of accretion discs and the innermost stable circular orbit.

Accretion discs are swirling disks of gas and dust that form around black holes, as well as other compact objects like neutron stars. They originate from a nearby star or a companion star in a binary system. As these stars evolve, they shed their outer layers, which then fall toward the black hole due to its immense gravitational pull. As the matter spirals inward, it forms a disc-like structure, heating up and emitting intense radiation before ultimately disappearing into the black hole's event horizon.

These discs provide a unique opportunity for astrophysicists to study the behavior of matter under extreme conditions. The intense gravitational forces and high temperatures within the accretion disc create a maelstrom of activity, generating powerful jets of particles and intense X-ray emissions. By observing these emissions, scientists can gain insights into the nature of black holes and the physics of the surrounding environment.

However, as matter spirals closer to the black hole, it encounters a critical point known as the innermost stable circular orbit (ISCO). This orbit represents the closest distance that matter can approach a black hole without falling into it. Beyond the ISCO, any matter is destined to be swallowed by the black hole's event horizon, disappearing from our observable universe forever.

Understanding the properties of the ISCO is crucial for unraveling the mysteries of black holes. The exact location of the ISCO depends on the mass and spin of the black hole, with more massive black holes having larger ISCO radii. Studying the ISCO can provide valuable information about the black hole's mass, spin, and the effects of general relativity.

Accretion discs and the innermost stable circular orbit offer a window into the fascinating world of black holes. They allow us to explore the extreme conditions that exist near these cosmic behemoths and shed light on the fundamental laws of the universe. Through meticulous observations and advanced theoretical models, astrophysicists continue to unlock the secrets of black holes, enhancing our understanding of the cosmos and our place within it.

Whether you are an avid astrophysics enthusiast or just someone curious about the wonders of the universe, delving into the intricacies of accretion discs and the innermost stable circular orbit will ignite your imagination and leave you in awe of the extraordinary phenomena that exist beyond our world.

Black Hole Spin and the Kerr Metric

In the vast expanse of the universe, black holes remain one of the most enigmatic and fascinating phenomena. These celestial objects, characterized by their immense gravitational pull, continue to captivate scientists and astronomers alike. One crucial aspect that contributes to the complexity of black holes is their spin, which is intricately connected to the Kerr Metric.

In the subchapter titled "Black Hole Spin and the Kerr Metric," we delve into the mysteries surrounding the rotation of black holes and its implications for our understanding of astrophysics. This content aims to provide a comprehensive introduction to the topic, catering to a diverse audience, including both newcomers and seasoned enthusiasts in astrophysics.

Firstly, we explore the concept of black hole spin, which refers to the rotational motion of these cosmic entities. Just like any other celestial object, black holes can also spin, and this rotation influences various aspects of their existence. We delve into the formation and evolution of black holes, shedding light on the mechanisms that lead to their spin.

To understand the significance of black hole spin, we introduce the Kerr Metric. This mathematical model, developed by physicist Roy Kerr, describes the spacetime around a rotating black hole. We explain how the Kerr Metric allows us to comprehend the complex interplay between gravity, rotation, and the curvature of spacetime near these extraordinary objects.

Furthermore, we discuss the consequences of black hole spin on their properties, such as their event horizon, ergosphere, and the intriguing phenomenon of frame-dragging. We explore how these factors affect the dynamics of matter and energy in the vicinity of a spinning black hole, and how they contribute to the formation of powerful astrophysical phenomena like relativistic jets.

Throughout this subchapter, we present visual aids, diagrams, and real-life examples to enhance the understanding of the concepts discussed. We also touch upon the current research and discoveries in the field, highlighting the ongoing efforts to unravel the secrets of black hole spin and its implications for our understanding of the universe.

By the end of this subchapter, readers will have gained a solid grasp of the intricate relationship between black hole spin and the Kerr Metric. They will appreciate the significance of this phenomenon in astrophysics and understand how it contributes to the extraordinary nature of black holes. Whether you are a novice or an expert in the field of astrophysics, this content will leave you with a deeper appreciation for the mysteries that lie within these cosmic behemoths.

Black Hole Entropy and the Information Paradox

In the vast and mysterious realm of astrophysics, few phenomena captivate the human imagination as much as black holes. These enigmatic cosmic entities, born from the death of massive stars, possess an immense gravitational pull that swallows everything in their path, including light itself. As we delve deeper into the study of black holes, we uncover an intriguing concept known as entropy, which sheds light on the complex nature of these cosmic enigmas and gives rise to the information paradox.

Entropy, often associated with disorder and chaos, is a fundamental concept in physics. It measures the level of randomness or uncertainty within a system. Surprisingly, black holes possess entropy, despite being characterized by their immense gravitational forces that seemingly annihilate everything that enters their event horizon. This revelation challenged the long-held notion that black holes were devoid of any characteristics, serving as mere cosmic vacuum cleaners.

The discovery of black hole entropy led to a profound realization: black holes contain information. This revelation sparked intense debate among astrophysicists, culminating in what is known as the information paradox. According to the laws of quantum mechanics, information cannot be destroyed. However, the immense gravitational forces within a black hole seemingly obliterate any information that falls within its grasp. This contradiction between the laws of quantum mechanics and the behavior of black holes poses a significant challenge to our understanding of the universe.

Various theories and hypotheses have been proposed to resolve the information paradox. One prominent idea suggests that information swallowed by a black hole is not lost but rather encoded in the black hole's event horizon. This concept, known as the holographic principle, postulates that the information is stored in a two-dimensional form on the surface of the event horizon, similar to a hologram. This theory provides a potential solution to the paradox, reconciling quantum mechanics with the behavior of black holes.

As we continue to unravel the secrets of black holes, the exploration of their entropy and the information paradox remains a fascinating frontier in astrophysics. Understanding how information behaves within the clutches of these cosmic behemoths could revolutionize our understanding of the universe and the fundamental laws that govern it.

Whether you are an aspiring astrophysicist or simply someone intrigued by the mysteries of the cosmos, the concept of black hole entropy and the information paradox invites you to embark on a mind-bending journey. Prepare to unlock the secrets of the universe as you delve into the depths of black holes, where entropy and information converge in a dance of cosmic proportions.

Chapter 6: Black Hole Phenomena and Effects

Hawking Radiation and Black Hole Evaporation

In the vast expanse of the cosmos, black holes have always captivated our imagination. These enigmatic cosmic entities, with their immense gravitational pull, have remained shrouded in mystery for centuries. However, in recent decades, scientists have made groundbreaking discoveries that have shed light on the secrets of black holes, including the mind-boggling concept of Hawking radiation and black hole evaporation.

First proposed by the brilliant physicist Stephen Hawking in 1974, Hawking radiation revolutionized our understanding of black holes. According to Hawking's theory, even though black holes are known for their ability to swallow everything, they are not completely black after all. In fact, they emit a faint radiation, now known as Hawking radiation, due to quantum effects near the event horizon – the point of no return.

But how does this radiation originate? According to quantum mechanics, a constant flux of virtual particles and antiparticles is constantly popping in and out of existence in the vacuum of space. Usually, these particle-antiparticle pairs annihilate each other almost instantaneously. However, near the event horizon of a black hole, one of these particles may fall into the black hole while its partner escapes, leading to the emission of Hawking radiation. This process results in the slow evaporation of black holes over time.

The discovery of Hawking radiation has profound implications for our understanding of the universe. It suggests that black holes are not eternal entities, destined to exist forever. Instead, they have a finite lifespan and eventually evaporate completely. For stellar-mass black holes, this process would take an unimaginably long time, trillions and trillions of years. However, for smaller black holes, the evaporation would be much faster, occurring within a fraction of a second.

The concept of black hole evaporation challenges the traditional view that nothing can escape a black hole's gravitational pull. Hawking radiation reveals that even black holes are subject to the laws of quantum physics. It also raises intriguing questions about the fate of information that falls into a black hole. As particles escape through Hawking radiation, it is still an open question as to whether the information they carry is preserved or lost forever.

The discovery of Hawking radiation and the concept of black hole evaporation have revolutionized our understanding of black holes, demonstrating the complex interplay between gravity and quantum mechanics. As we continue to delve deeper into the mysteries of the cosmos, these revelations bring us one step closer to unlocking the secrets of black holes and unraveling the fundamental nature of our universe.

This subchapter explores the revolutionary concept of Hawking radiation and black hole evaporation, revealing the dynamic and evolving nature of these celestial phenomena. Whether you are an astrophysics enthusiast or simply curious about the mysteries of the universe, this chapter will take you on a journey into the fascinating world of black holes and the profound impact they have on our understanding of the cosmos.

Active Galactic Nuclei and Quasars

In the vast expanse of the universe, there exist celestial objects that defy our understanding and captivate the imagination of astrophysicists and enthusiasts alike. Among these enigmatic structures are Active Galactic Nuclei (AGNs) and their highly energetic counterparts, quasars. These cosmic powerhouses are some of the most intriguing phenomena in astrophysics, and their study has opened up new frontiers in our quest to understand the mysteries of the universe.

An AGN refers to the central region of a galaxy that emits an extraordinary amount of energy across the entire electromagnetic spectrum. This emission arises due to the presence of a supermassive black hole at the heart of the galaxy, surrounded by a swirling disk of gas and dust known as an accretion disk. As matter falls into the black hole, it forms a hot, glowing disk that releases immense energy in the form of radiation, making AGNs some of the brightest objects in the universe.

Quasars, short for "quasi-stellar radio sources," are a subset of AGNs that emit colossal amounts of energy. They are characterized by their extreme brightness and ability to outshine an entire galaxy. Quasars are thought to be powered by the same mechanisms as AGNs, but they appear much brighter due to their orientation towards Earth, allowing us to observe them as highly luminous objects across vast distances.

The study of AGNs and quasars not only provides insights into the nature of supermassive black holes but also offers a glimpse into the evolution of galaxies. By analyzing the emission lines in their spectra, astrophysicists can determine the composition of the surrounding gas and understand the physical processes occurring in the vicinity of these cosmic powerhouses. This knowledge helps us unravel the intricate interplay between the black hole and its host galaxy, shedding light on the mechanisms that drive galaxy formation and evolution.

Furthermore, studying AGNs and quasars provides a unique opportunity to probe the fundamental physics governing the universe. The extreme conditions near the event horizon of a supermassive black hole allow scientists to test the limits of our current understanding of gravity, spacetime, and the behavior of matter under extreme gravitational forces.

As we continue to unlock the secrets of black holes, the study of AGNs and quasars remains a crucial area of investigation. The discoveries made in this field not only push the boundaries of astrophysics but also deepen our understanding of the universe we inhabit. By unraveling the mysteries of these cosmic powerhouses, we embark on a journey that brings us closer to comprehending the vast complexities of the cosmos.

Black Hole Jets and Relativistic Outflows

In the vast and mysterious realm of astrophysics, black holes are undoubtedly one of the most captivating phenomena. These cosmic enigmas possess such immense gravitational forces that not even light can escape their clutches, making their study a truly mind-bending endeavor. However, the secrets of black holes extend far beyond their event horizons. One of the most fascinating aspects of these cosmic monsters is their ability to generate powerful jets and relativistic outflows.

Imagine, if you will, a black hole devouring matter from its surroundings. As matter spirals into the black hole's gravitational grasp, it forms an accretion disk around it. This disk consists of superheated gas and dust particles swirling at incredible speeds. Within this swirling maelstrom, a fraction of particles is accelerated to relativistic speeds, forming high-energy jets that shoot out perpendicular to the accretion disk.

These jets are not only visually striking but also immensely powerful. They can extend for thousands or even millions of light-years, carrying vast amounts of energy with them. Although scientists have yet to fully comprehend the mechanisms behind jet formation, it is believed that they are powered by the immense rotational energy of the black hole itself.

Relativistic outflows, on the other hand, are a broader term encompassing a range of phenomena associated with black holes. These outflows can take various forms, including powerful winds, particle beams, and even gamma-ray bursts. They occur when the gravitational energy of the black hole is converted into kinetic energy, propelling particles and radiation away from the vicinity of the black hole.

The study of black hole jets and relativistic outflows provides invaluable insights into the fundamental physics governing these enigmatic objects. By analyzing the composition, speed, and energy of these phenomena, astrophysicists can unravel the mysteries of how black holes form, grow, and impact their surrounding environments. Moreover, understanding these processes can shed light on the broader astrophysical phenomena, such as galaxy formation and evolution, as black holes are intricately linked to the cosmic web of matter and energy.

While the study of black hole jets and relativistic outflows is undoubtedly complex, it is also a testament to the human quest for knowledge and understanding. Each discovery made in this field brings us closer to unlocking the secrets of the universe and comprehending the awe-inspiring forces that shape it. So, whether you are an aspiring astrophysicist or simply a curious mind, delving into the world of black hole jets and relativistic outflows promises to be an extraordinary journey that will expand your cosmic horizons.

Chapter 7: Black Holes and the Universe

Black Holes as Galaxy Builders

In the vast expanse of the universe, black holes have long captivated the imagination of scientists and the general public alike. Often portrayed as destructive and mysterious entities, their true nature and role in shaping the cosmos have remained enigmatic. However, recent discoveries have shed light on a remarkable aspect of black holes - their ability to act as galaxy builders.

A galaxy is a colossal assembly of stars, gas, dust, and dark matter, held together by the force of gravity. These cosmic islands come in various shapes and sizes, from spirals to ellipticals, and understanding their formation is crucial to unraveling the secrets of the universe. It turns out that black holes, with their immense gravitational pull, play a pivotal role in this process.

At the heart of most galaxies lies a supermassive black hole, millions or even billions of times more massive than our Sun. These behemoths are formed through the accumulation of matter over billions of years. As matter falls into the black hole, it forms an accretion disk, a swirling ring of gas and dust. During this process, particles collide and release energy, emitting intense radiation that can be detected across the electromagnetic spectrum.

This radiation can have profound effects on the surrounding galaxy. The energy emitted from the accretion disk heats up the gas and dust, triggering the birth of new stars. It also drives powerful outflows of gas, creating galactic winds that sweep across vast distances and influence the formation of neighboring galaxies. In this way, black holes act as cosmic architects, sculpting the structure and evolution of galaxies.

Furthermore, black holes are not only responsible for galaxy formation but also for maintaining their delicate balance. As matter falls into the black hole, it releases enormous amounts of energy, which can regulate the growth of the galaxy. This process, known as feedback, prevents excessive star formation and stabilizes the galaxy over long periods of time.

The discovery of black holes as galaxy builders has revolutionized our understanding of astrophysics. It highlights the intricate interplay between these enigmatic objects and the cosmic structures they inhabit. By unraveling the secrets of black holes, scientists are unlocking the mysteries of the universe itself.

In conclusion, black holes are not just cosmic monsters lurking in the depths of space; they are essential forces driving the formation and evolution of galaxies. Their enormous gravitational pull and energetic emissions shape the cosmic landscape, triggering the birth of stars and influencing the growth of neighboring galaxies. Understanding the role of black holes as galaxy builders is a crucial step in unraveling the secrets of the universe and advancing our knowledge of astrophysics.

Black Holes and Dark Matter

In the vast expanse of the universe, there exist enigmatic cosmic entities that continue to captivate and challenge the minds of astrophysicists: black holes and dark matter. The mysteries surrounding these phenomena have puzzled scientists for decades, prompting a relentless pursuit to unlock their secrets and understand their profound implications.

Black holes, often referred to as the ultimate cosmic abyss, are regions in space with such immense gravitational force that nothing, not even light, can escape their grasp. These celestial objects are formed when massive stars collapse under their own weight after exhausting their nuclear fuel. The resulting singularity, a point of infinite density, is surrounded by an event horizon, a boundary beyond which nothing can return.

One of the most intriguing aspects of black holes is their ability to warp the very fabric of space and time. This distortion creates a gravitational pull so strong that it can bend light rays, leading to phenomena like gravitational lensing. Through the study of black holes, scientists hope to gain a deeper understanding of the nature of gravity and the fundamental laws that govern the universe.

Dark matter, on the other hand, presents a different kind of mystery. It is a form of matter that does not interact with light or electromagnetic radiation, making it invisible to our traditional methods of observation. Yet, its presence can be inferred through its gravitational effects on visible matter. In fact, it is estimated that dark matter accounts for a significant portion of the universe's mass, far outweighing the ordinary matter we are familiar with.

Understanding dark matter is crucial to unraveling the structure and evolution of galaxies. Its gravitational pull helps hold galaxies together, preventing them from dispersing into the vastness of space. Scientists believe that dark matter played a vital role in the formation of galaxies and the large-scale structures we observe today.

While much about black holes and dark matter remains shrouded in mystery, advances in astrophysics have enabled us to glimpse into their secrets. Through the use of powerful telescopes, such as the Hubble Space Telescope and the Event Horizon Telescope, scientists have made groundbreaking discoveries, providing valuable insights into these cosmic enigmas.

As we delve deeper into the realms of astrophysics, the exploration of black holes and dark matter continues to push the boundaries of our knowledge. By unraveling the mysteries they hold, we not only gain a better understanding of the universe but also glimpse the infinite possibilities that lie beyond the limits of our perception.

The Role of Black Holes in Cosmic Evolution

In the vast expanse of the universe, black holes have long captivated the minds of scientists, artists, and enthusiasts alike. These enigmatic cosmic entities have been a subject of fascination and speculation for centuries. However, it is only in recent years that astrophysics has begun to unravel the secrets they hold and understand their pivotal role in cosmic evolution.

Black holes, as the name suggests, are regions in space with an incredibly strong gravitational pull that nothing, not even light, can escape. They form from the remnants of massive stars that have exhausted their nuclear fuel and collapse under their own gravity. Once formed, black holes continue to grow by devouring surrounding matter, including stars, planets, and even other black holes.

One of the most significant contributions of black holes to cosmic evolution is their role in shaping galaxies. Galaxies are collections of billions of stars held together by gravity, and black holes reside at their centers. These supermassive black holes, with masses millions or even billions of times that of our sun, have a profound impact on the growth and structure of galaxies.

As matter falls into a black hole, it forms a rotating disk known as an accretion disk. The extreme gravitational forces and intense friction within this disk generate tremendous amounts of energy, releasing powerful jets of particles and radiation into space. These jets can influence the formation of new stars, triggering the birth of vast stellar nurseries within galaxies.

Furthermore, the immense gravity of black holes can create a phenomenon known as gravitational lensing. As light passes near a black hole, its path is bent, causing distant objects to appear magnified or distorted. This effect allows astronomers to study galaxies and other celestial objects that would otherwise be invisible or obscured by interstellar dust.

Black holes also play a crucial role in the life cycle of stars. When a massive star exhausts its nuclear fuel, it can undergo a supernova explosion, leaving behind a dense core known as a neutron star or, in some cases, collapsing into a black hole. The energy released during this process can spur the formation of new stars, continuing the cycle of stellar birth and death.

Understanding the properties and behaviors of black holes is not only an intellectual pursuit but also a vital component of astrophysics. By studying these cosmic powerhouses, scientists can gain insights into the fundamental laws of nature, the origins of the universe, and the intricacies of space-time itself.

In conclusion, black holes are not just cosmic oddities; they are key players in the grand drama of cosmic evolution. From shaping galaxies to influencing star formation, their immense gravity and unique properties have a profound impact on the universe as we know it. As we continue to unlock the secrets of black holes, we uncover a deeper understanding of our place in the cosmos and the forces that have shaped our existence.

Chapter 8: The Future of Black Hole Astrophysics

Advancements in Technology and Observational Techniques

The field of astrophysics has undergone remarkable advancements in recent years, thanks to the rapid progress in technology and observational techniques. These developments have allowed scientists to delve deeper into the mysteries of the universe, particularly in understanding the enigmatic phenomena of black holes. In this subchapter, we will explore the groundbreaking innovations that have propelled astrophysics forward and revolutionized our understanding of black holes.

One of the key technological advancements is the construction of sophisticated telescopes. The advent of space-based observatories, such as the Hubble Space Telescope and the Chandra X-ray Observatory, has provided astronomers with unprecedented views of the cosmos. These telescopes capture images and data across various wavelengths, enabling scientists to observe black holes and their surroundings with remarkable clarity. By combining data from different telescopes, scientists can create a comprehensive picture of these cosmic beasts.

Furthermore, the development of gravitational wave detectors has opened up an entirely new avenue for studying black holes. Gravitational waves, ripples in the fabric of spacetime caused by cataclysmic events, allow us to directly detect black hole mergers and study their properties. The Laser Interferometer Gravitational-Wave Observatory (LIGO) was the first to successfully detect gravitational waves in 2015, marking a monumental achievement in astrophysics. Since then, additional detectors have been built, expanding our ability to observe these elusive phenomena.

Advancements in computer simulations and modeling techniques have also played a crucial role in studying black holes. Supercomputers can now simulate the behavior and evolution of black holes, providing valuable insights into their formation, growth, and interactions with surrounding matter. These simulations help scientists test theoretical models and make predictions about the behavior of black holes under various conditions.

Moreover, the advent of big data analytics and machine learning algorithms has significantly enhanced our ability to process and analyze vast amounts of astronomical data. These techniques allow scientists to sift through enormous datasets and identify patterns or anomalies that may provide clues about the nature of black holes.

In conclusion, the advancements in technology and observational techniques have revolutionized the field of astrophysics, particularly in our understanding of black holes. The construction of advanced telescopes, the detection of gravitational waves, the development of computer simulations, and the utilization of big data analytics have all contributed to unlocking the secrets hidden within these cosmic wonders. As technology continues to advance, we can expect even more astonishing discoveries and a deeper understanding of the enigmatic black holes that populate our universe.

Unanswered Questions and Areas of Research

Astrophysics, the study of celestial objects and the universe at large, has always been a field of endless fascination and exploration. While significant progress has been made in understanding the mysteries of the cosmos, there are still numerous unanswered questions that continue to fuel the curiosity of scientists and enthusiasts alike. In this subchapter, we delve into some of these lingering enigmas and the exciting areas of research that are pushing the boundaries of our knowledge.

One of the most perplexing phenomena in astrophysics is the enigma of black holes. Despite recent breakthroughs, there are still fundamental questions surrounding their formation, behavior, and ultimate fate. Researchers are actively investigating the mechanisms by which black holes are created, whether through the collapse of massive stars or other unknown processes. Understanding the intricate dynamics within a black hole's event horizon remains a tantalizing challenge, as it defies our current understanding of the laws of physics.

Another area of intense research is the nature of dark matter and dark energy, which together constitute a staggering 95% of the universe. Although their existence is inferred from the gravitational effects they exert, their true identities remain elusive. Scientists are developing novel experimental techniques and theoretical models to probe the properties of these enigmatic substances. Unlocking the secrets of dark matter and dark energy could revolutionize our understanding of the universe, shedding light on its structure, evolution, and ultimate fate.

Furthermore, the origin of cosmic rays, high-energy particles that bombard our planet from outer space, is yet to be fully explained. Astrophysicists are investigating potential sources, such as supernovae, pulsars, and even supermassive black holes, to unravel the mysteries behind these energetic messengers. Understanding the composition, acceleration mechanisms, and distribution of cosmic rays could provide vital insights into the cosmic processes that shape our universe.

Lastly, the search for extraterrestrial life remains an enduring question in astrophysics. Scientists are actively exploring the possibilities of habitable environments beyond Earth, such as exoplanets within the habitable zone of their star systems. The discovery of potentially habitable exoplanets and the ongoing development of advanced telescopes and detection methods offer promising avenues for detecting signs of life elsewhere in the universe.

As we continue to unlock the secrets of black holes and delve deeper into the mysteries of the universe, it is important to remember that astrophysics is not solely the domain of scientists. The wonders of the cosmos are open to everyone, inviting curiosity, awe, and a thirst for knowledge. By nurturing our collective fascination and supporting ongoing research, we can strive to answer these unanswered questions and expand our understanding of the awe-inspiring universe in which we live.

Theoretical Models and the Quest for a Unified Theory

In the vast realm of astrophysics, scientists are constantly striving to unravel the mysteries of the universe. One of the most intriguing and enigmatic phenomena that has captivated the minds of researchers for centuries is the black hole. These cosmic entities, with their immense gravitational pull, have intrigued scientists and sparked numerous theoretical models in an attempt to understand their nature.

The quest for a unified theory is at the heart of the scientific community's efforts to comprehend black holes. A unified theory aims to merge the laws of physics, particularly Einstein's theory of general relativity and quantum mechanics, into a single framework that can explain all aspects of the universe. However, this endeavor has proven to be a formidable challenge, as the laws governing the macroscopic world of gravity and the microscopic world of quantum mechanics seem to clash.

Various theoretical models have been proposed to bridge this gap and provide a unified explanation for black holes. One such model is string theory, which suggests that the fundamental building blocks of the universe are tiny, vibrating strings. These strings exist in multiple dimensions, and their different vibration modes correspond to different particles and forces in the universe. String theory offers a potential solution to the inconsistencies between general relativity and quantum mechanics, providing a promising avenue for understanding the physics of black holes.

Another theoretical framework that has gained prominence in recent years is loop quantum gravity. This model approaches the problem from a different angle, focusing on the discrete nature of space and time. Loop quantum gravity suggests that space is made up of tiny, indivisible units or "loops." By quantizing space and time, this theory offers a new perspective on the extreme conditions within black holes, potentially revealing the secrets hidden within their event horizons.

While these theoretical models provide valuable insights into the nature of black holes, the quest for a unified theory is ongoing. The scientific community continues to explore new ideas, test hypotheses, and refine existing models in the pursuit of a comprehensive understanding of these cosmic enigmas.

For anyone intrigued by the mysteries of the universe, the quest for a unified theory represents a fascinating journey into astrophysics. By delving into theoretical models and their implications for black holes, we gain a deeper appreciation for the complexity of the cosmos and the relentless pursuit of knowledge that drives scientists forward. As our understanding of the universe expands, we come closer to unlocking the secrets of black holes and unraveling the mysteries of the universe.

Chapter 9: The Impact of Black Hole Astrophysics on Society

Popular Culture and Black Hole Misconceptions

In today's age of technology and media, popular culture has a significant influence on how we perceive and understand scientific concepts. One such concept that has captured the imagination of both scientists and the general public is the enigmatic phenomenon known as black holes. These celestial objects, with their immense gravitational pull, have been portrayed in movies, books, and even music, often leading to misconceptions and misunderstandings about their true nature.

When it comes to black holes, popular culture often portrays them as menacing, all-consuming monsters that can suck in everything around them, even light itself. While it is true that black holes have a powerful gravitational force, it is essential to understand that they do not suck in everything indiscriminately. Black holes are formed from massive stars that have collapsed under their own gravity, leaving behind an incredibly dense core. Their gravitational pull is only felt within a certain distance, known as the event horizon, beyond which nothing can escape, not even light. The misconception arises when popular culture exaggerates this effect, leading to the belief that black holes can swallow entire galaxies or devour spaceships in a matter of seconds.

Another common misconception perpetuated by popular culture is the idea that black holes are "holes" in space, leading to other dimensions or parallel universes. While the term "hole" is used metaphorically, it can be misleading. In reality, black holes are not portals or gateways to other realms. They are singularities, points of infinite density, where the laws of physics, as we currently understand them, break down. The concept of what lies beyond a black hole's event horizon remains a mystery, and scientists are still exploring theories and possibilities in this regard.

It is crucial to differentiate between the scientific understanding of black holes and the portrayals in popular culture. While movies like Interstellar and books like Stephen Hawking's A Brief History of Time have made efforts to depict black holes more accurately, the sensationalized and exaggerated versions found in other media can lead to misconceptions and misunderstandings.

As astrophysics enthusiasts, it is our responsibility to critically analyze and question the information we receive from popular culture. By seeking out reliable sources and engaging with scientific literature, we can separate fact from fiction and deepen our understanding of black holes. Only through education and accurate information can we unlock the secrets of these fascinating cosmic entities and truly comprehend their place in the vast expanse of the universe.

Practical Applications of Black Hole Knowledge

Black holes have long captivated the imaginations of scientists and the public alike. These mysterious cosmic entities, with their intense gravitational pull and ability to trap even light, have been the subject of extensive research in the field of astrophysics. While the study of black holes is undoubtedly fascinating, it may seem purely theoretical to some. However, recent advancements in our understanding of black holes have given rise to exciting practical applications that can impact various aspects of our lives.

One of the most significant practical applications of black hole knowledge lies in the field of space exploration. Black holes are known to warp the fabric of spacetime, offering a potential means of faster-than-light travel. By harnessing the gravitational forces near a black hole, scientists have proposed the concept of wormholes, hypothetical shortcuts that could connect distant parts of the universe. While the practical implementation of such wormholes remains a challenge, the study of black holes has opened new possibilities for interstellar travel and the exploration of distant galaxies.

Moreover, black holes have a crucial role in understanding the formation and evolution of galaxies. Galaxies, including our own Milky Way, have supermassive black holes at their centers. By studying the behavior of matter and energy around these black holes, astrophysicists gain insights into the processes that shape galaxies. This knowledge not only enhances our understanding of the universe but also contributes to advancements in cosmology and astrophysical simulations.

In addition to space exploration and galaxy formation, black hole knowledge has found practical applications on a smaller scale as well.

Theoretical models based on black hole physics have aided in the development of advanced imaging techniques, such as medical imaging technologies like CT scans and MRI machines. These imaging methods rely on the principles of black hole physics to capture detailed images of internal structures, enabling accurate diagnoses and improved medical treatments.

Furthermore, black holes play a vital role in the field of gravitational wave astronomy. When massive objects, such as black holes or neutron stars, collide, they generate ripples in spacetime known as gravitational waves. Detecting and analyzing these waves provide valuable insights into the nature of black holes, as well as the fundamental properties of the universe. This emerging field of research holds great potential for technological advancements, including the development of more accurate and sensitive detectors for various applications.

In conclusion, the practical applications of black hole knowledge extend far beyond the realms of astrophysics. From space exploration and galaxy formation to medical imaging and gravitational wave astronomy, our understanding of black holes has opened doors to new frontiers of discovery and innovation. As we continue to unravel the secrets of these enigmatic cosmic entities, the practical benefits derived from this knowledge will undoubtedly contribute to advancements in various scientific fields and impact our lives in profound ways.

Ethical Considerations in Black Hole Research

As humanity continues to unravel the mysteries of the universe, astrophysics stands at the forefront of scientific exploration. One of the most captivating and enigmatic entities in the cosmos is the black hole. These cosmic behemoths possess immense gravitational forces that can warp space and time, captivating both scientists and the public alike. However, delving into the depths of black hole research raises profound ethical considerations that must be addressed.

First and foremost, the study of black holes necessitates the use of advanced technology and resources. The allocation of these resources raises questions about the ethical implications of prioritizing astrophysics research over other areas of societal need. As we venture into the depths of space, we must ensure that our pursuit of knowledge does not neglect pressing issues on our own planet, such as poverty, education, or healthcare.

Furthermore, the collection of data for black hole research often relies on the observation of celestial bodies and phenomena. This raises concerns about the potential invasion of privacy for distant objects in space. While it may seem trivial to consider the privacy rights of inanimate objects, it is crucial to adopt an ethical framework that respects the autonomy and dignity of all entities, regardless of their physical nature.

Another key ethical consideration lies in the potential consequences of black hole research. As we strive to understand these cosmic wonders, there is a possibility that our experiments could inadvertently cause harm. For instance, attempts to manipulate black holes or artificially create mini black holes in laboratories may have unforeseen consequences for the fabric of space-time or even the stability of our own planet. Therefore, it is paramount to exercise caution and implement rigorous safety protocols to mitigate any potential risks.

Additionally, the dissemination of knowledge gained from black hole research raises ethical questions about access and inclusivity. As scientists make groundbreaking discoveries, it is imperative to ensure that this knowledge is accessible to everyone, regardless of their socioeconomic background or geographic location. Open access to scientific findings promotes a more equitable society and enables individuals from diverse backgrounds to contribute to the advancement of astrophysics.

In conclusion, while the study of black holes offers immense possibilities for scientific advancement, it is crucial to approach this research with careful ethical considerations. Balancing the allocation of resources, respecting privacy rights, mitigating potential risks, and promoting inclusivity are all integral aspects of responsible black hole exploration. By navigating these ethical challenges with diligence and integrity, we can unlock the secrets of black holes while upholding our moral responsibilities to society and the universe at large.

Chapter 10: Conclusion: Our Journey into Black Hole Astrophysics

Recap of Key Insights and Discoveries

In the ever-expanding field of astrophysics, our understanding of black holes has undergone a remarkable transformation. Throughout this journey of exploration and discovery, we have unraveled some of the most intriguing secrets of these enigmatic cosmic entities. In this subchapter, we will recapitulate the key insights and discoveries that have revolutionized our understanding of black holes.

First and foremost, we have come to understand that black holes are not merely celestial objects with an extraordinarily strong gravitational pull. They are formed from the remnants of massive stars that have exhausted their nuclear fuel and undergone a catastrophic collapse. This collapse results in the formation of a singularity, a point of infinite density at the center of a black hole, surrounded by an event horizon beyond which nothing can escape its gravitational grasp.

One of the most groundbreaking discoveries in recent years is the evidence for the existence of supermassive black holes at the centers of galaxies. These colossal black holes, millions or even billions of times more massive than our Sun, play a pivotal role in shaping the evolution of galaxies. They have been found to influence the motion of stars and gas within their vicinity and are responsible for the formation of active galactic nuclei and quasars, some of the most energetic phenomena in the universe.

Another significant insight is the discovery of gravitational waves, ripples in the fabric of spacetime, which were first predicted by Einstein's theory of general relativity. In 2015, the Laser Interferometer

Gravitational-Wave Observatory (LIGO) made the historic detection of gravitational waves originating from the merger of two black holes. This groundbreaking observation not only confirmed Einstein's theory but also opened up an entirely new window through which we can study the universe.

Furthermore, recent studies have shed light on the intriguing concept of black hole evaporation. According to physicist Stephen Hawking, black holes are not completely black; they emit a faint radiation known as Hawking radiation. This phenomenon suggests that black holes slowly lose mass over time and eventually evaporate completely, leaving behind no trace of their existence.

In conclusion, the field of astrophysics has witnessed remarkable progress in unraveling the secrets of black holes. From understanding their formation and role in galaxy evolution to detecting gravitational waves and contemplating their eventual demise, our journey into the realm of black holes has been nothing short of extraordinary. As we continue to delve deeper into this captivating field, it is certain that more revelations and discoveries await, expanding our knowledge of the universe and challenging our perceptions of reality.

Inspiring the Next Generation of Astrophysicists

Astrophysics, the branch of science that unravels the mysteries of the universe, has always fascinated both young and old alike. From the breathtaking beauty of distant galaxies to the mind-boggling concept of black holes, the study of astrophysics has the power to ignite curiosity and inspire a sense of wonder within us. In this subchapter, we delve into the importance of inspiring the next generation of astrophysicists and nurturing their passion for this awe-inspiring field.

Astrophysics is not just an academic pursuit for a select few; it is a subject that can captivate the imagination of everyone, regardless of age or background. By exposing children and young adults to the wonders of astrophysics, we can foster a generation of thinkers and problem solvers who will shape the future of this fascinating field. The key lies in providing accessible and engaging opportunities for them to explore the cosmos.

One way to inspire the next generation of astrophysicists is through educational outreach programs. By partnering with schools and community organizations, astrophysicists can bring the wonders of the universe directly to students. From interactive stargazing sessions to hands-on experiments and demonstrations, these programs can spark a sense of awe and curiosity, planting the seeds of a lifelong interest in astrophysics.

Another avenue for inspiration is through mentorship and role models. Young minds are often influenced by those who have paved the way before them. By showcasing the achievements and stories of diverse astrophysicists, we can inspire underrepresented groups and break down barriers that may have previously existed. Encouraging mentorship programs and providing opportunities for aspiring young astrophysicists to connect with professionals in the field can be instrumental in fostering their passion and keeping them engaged in their journey.

In addition to outreach and mentorship, it is vital to make astrophysics accessible to everyone, regardless of their level of scientific expertise. By using plain language and avoiding jargon, we can bridge the gap between complex astrophysical concepts and a wider audience. Engaging books, documentaries, podcasts, and even online courses can serve as valuable tools to ignite curiosity and inspire further exploration.

In conclusion, inspiring the next generation of astrophysicists is a crucial endeavor. By reaching out to students, providing mentorship, and making astrophysics accessible, we can nurture their curiosity and passion for the cosmos. Together, let us unlock the secrets of black holes and embark on a journey into astrophysics that transcends boundaries and inspires generations to come.

Final Thoughts and Contemplations

As we conclude this fascinating journey into the depths of astrophysics and the enigmatic world of black holes, it is important to reflect on the incredible knowledge we have gained. Throughout this book, we have explored the mysteries of the universe, delving into the awe-inspiring phenomena that black holes represent. Now, let us take a moment to contemplate the significance of our discoveries and their implications for both the field of astrophysics and humanity as a whole.

One cannot help but be humbled by the sheer magnificence of black holes. These celestial objects, with their immense gravitational pull, have the power to warp space and time, confounding our understanding of the universe. They challenge our notions of physics and stretch the limits of our imagination. Studying black holes has opened up new avenues of research, prompting us to question the fundamental laws that govern our reality.

In the realm of astrophysics, the study of black holes has revolutionized our understanding of the cosmos. We now know that these cosmic behemoths are not merely theoretical constructs but are present throughout the universe. They play a vital role in the evolution of galaxies, shaping the very fabric of space on cosmic scales. Unraveling the secrets of black holes has provided us with invaluable insights into the nature of gravity, the birth and death of stars, and the evolution of galaxies.

But the impact of black holes extends far beyond the realm of astrophysics. They captivate the public's imagination and spark curiosity about our place in the universe. They inspire awe and wonder, reminding us of the vastness and complexity of the cosmos. Black holes have become cultural icons, appearing in literature, movies, and art, serving as metaphors for the unknown and the mysterious.

As we conclude this journey, let us remember that the exploration of black holes is an ongoing endeavor. Our understanding is constantly evolving as we push the boundaries of knowledge. Unlocking the secrets of black holes requires collaboration, innovation, and a relentless pursuit of the truth.

So, whether you are an astrophysics enthusiast or simply curious about the wonders of the universe, may this book have ignited a spark within you. May it have inspired you to delve deeper into the mysteries of black holes and to ponder the profound questions they pose. The journey into astrophysics is an ever-unfolding odyssey, and by embracing the enigma of black holes, we embark on a quest that will continue to shape our understanding of the cosmos for generations to come.

www.ingramcontent.com/pod-product-compliance
Lightning Source LLC
Chambersburg PA
CBHW051354150726
48000CB00003B/1179